The 10-SECONDS

SPEED

MATHS

TECHNIQUE

The Guaranteed Path to Competence and Enjoyment in Maths

By

G000065406

PHILIP CHAN

www.the10-secondsmathsexpert.com

First edition published in USA by 10-10-10 Publishing.

Second edition published in UK by DVG STAR Publishing

Designs and Patents Act 1988 to be identified as the author of this work.

Book cover design by SWISSCREATIONS

Copyright © 2016 Philip Chan

All rights reserved.

ISBN: 0992869447
ISBN-13: 978-0992869441

DEDICATION

This book is dedicated to You the Reader.

The knowledge that I share with you here will enable you to take control of your mind, body and lifestyle. Using the experience gained from my own struggles in life, I have outlined some of the powerful techniques that you can follow to overcome the challenges and conquer the fears of doing mathematics.

The key to achieving life-changing results is to understand, educate, empower and encourage yourself to take inspired action. The aim of this book is to inspire you to eliminate all traces of self-doubt, to think positively and act with confidence.

How?

By working with the elements in your life which influence your mental and physical wellbeing, I will you give you a glimpse of how you can take control of your life.

Success is defined by your ability to find your purpose in life.

"Your only limitation is what you have put on yourself!"

Author's notes

For Readers in USA and Canada

The abbreviations for 'Mathematics' in UK is 'Maths'

Instead of 'Math' in USA and Canada

Author's notes on Second Edition

It is with grateful thanks to the many readers from the First Edition that I have included some extra materials in this edition to enhance your learning.

Have fun and enjoy this edition if you are either reading this work again or a new reader.

CONTENTS

ACKNOWLEDGMENTS

If it were not for Raymond Aaron introducing me to his powerful 10-10-10 program and the wonderful guidance from Lori Murphy, my Book Architect, this book would never have gotten passed the starting block.

I am also grateful to all the wonderful support and encouragement from the community of

10-10-10 program writers, particularly to Marina Nani for her believe in all of us to shine.

There have been times I could have easily given up if it was not for the continuing cheering on by a few good friends like John and Beverley Neilson, my lifelong friends and mentors; Terence Andrews , a close friend and the most inspirational mathematics teacher I know ; Ahmed El-Amine , a personal friend and fellow mathematician who is constantly encouraging me to focus to do the right things; Paul Treanor, my personal M.A.T. coach helping me restoring the mobility of my legs which I am eternally grateful..

A special mention and sincere grateful thanks to Joe Gregory (Co-author of 'The Wealthy Author' with Debbie Jenkins) for his invaluable tips and advice for my first book.

I also want to thank my mentor J T Foxx (The World's No 1 Coach and Business Strategist) for his inspirational teaching and sharing his incredible journey of transformation from adversity to one of the most dynamic global business leaders of the Modern Era. He is my business Guru.

I want to thank my brother Alfred Chan for all his assistance in doing some of the graphics, without whom I would have been totally lost, and my sister Linda Chan for taking care of family matters to allow me the time to write this book. She is a wonderful teacher and Dance Therapist in her own right.

This book is a combination of the influences of hundreds of ideas from great mathematicians down the ages and during my teaching career from other mathematics professionals.

A special mention must be made to the originator idea of the 3-6-9 times table, Thomas Biesanz and his wonderful book called 'Right Brain Maths'. You can find out more of his wonderful work on http://RightBrainMath.com.

I wanted to leave behind a legacy for some very special people in my life:

Josh and Fynn, Whiting, Kristie & Ben Welch, Freya Whiting, Leah, Moya, Chloe and Jack Whiting. I hope that one day each one of them will make their contributions in the world and share their stories.

Finally, I want to thank all of my former pupils, parents and teachers over the past 40 years who I have worked with and who attended my workshops. They gave me the chance to refine my skills and share the many techniques I have discovered in my research by the great Mathematicians throughout the ages and from different cultures for their work to enrich us all.

I have a request of you. When you learn something new, please go and share it.

An old Swedish proverb: "A problems shared is a problem halved. Shared joy – Double joy!"

Thank you for purchasing this book and please always share the fun and knowledge with others.

All the success and best wishes.

Philip Chan

The 10-Seconds Maths Expert

.

FOREWORD

'The 10-Seconds Speed Maths Technique' by Philip Chan is intended to help you the reader to have fun with your children to achieve mega success in learning maths together.

By learning the simple, powerful and stimulating techniques outlined in his book, it will empower you the reader to learn quickly and easily the basic in maths in order to achieve instant success.

In this book:

- You will discover how to learn your individual times tables in less than a minute.

- You will amaze yourself and your friends by how quickly you can do calculations either mentally or on paper in a fraction of the time.

- You will develop an important part of your brain known as RAS
  (Reticular Activated System) being aware of patterns which were previously overlooked.

- You will learn how to utilise the key skills in your brain development in other areas of learning.

- You will see the growth of confidence in your children as they practise the skills in this book.

- You will start to develop the confidence and power in the 'I CAN' spirit for successful learning.
- This book will empower you the reader to challenge your beliefs to find the inner power for greater achievement and success.

Raymond Aaron

NY Times Bestselling Author

www.UltimateAuthorBootcamp.com

FEEDBACK AND TESTIMONIALS

Testimonial from Teachers and Professionals

Having attended Philip Chan's live workshops with my own child, I know that a huge number of children (and parents) have already benefitted from his infectious enthusiasm and memorable

'The 10- Seconds Speed Maths Technique'. I am really pleased his work is now more widely available through this book. Enjoy!

-Kate Bellingham (former BBC Television presenter on 'Tomorrow's World')

Philip's natural intelligence, empathy and dynamism make him the most gifted educator I have ever seen in over the past 35 years

– David Akerman (Chamber of Commerce and Government funding bodies for over 25 years)

The volume, enthusiasm respect of his students for Philip Chan's method of teaching is held in tremendously high regard as 'The 10-Seconds Maths Wizard' by many teachers and pupils alike!

–Terry Andrews (Former County Maths Consultant)

I was introduced to Philip Chan back in the year 2000 whilst training as a Maths teacher. I observed Philip's teaching his fun and innovative techniques with students across the age and ability ranges and was amazed at the simplicity and effectiveness. I subsequently attended Philip's popular live workshops to learn more of 'The 10-Seconds Speed Maths Technique'. These have served me well throughout my teaching career and I am thrilled to see them published in a book, although I hope this is only the first of many."

-David Mills (Maths & Professional Confidence NLP coach)

Philip Chan….the hub of knowledge and his method inspire us Maths professional with

a 'Can Do' attitude to pass on to our students.

-Ahmed El Amine (Maths Teacher & Numeracy Coordinator)

FEEDBACK FROM STUDENTS AND PARENTS

My daughter and I attended a number of maths sessions run by Philip Chan and we both came away with a rekindled desire for maths which is still burning inside us today.

- Mary Heffernan (Parent)

My daughters were struggling with their maths until with Philip's expert tuition and 'The 10-Seconds Speed Maths Technique' restored their confidence instantly. Using his methods my eldest daughter improved her GCSE grades from Grade D to A and my youngest from E to C. I cannot thank him enough for sharing these wonderful stress-free and fun methods.

- Nicola Bird (Parent)

I loved the way Philip Chan helped us all our times tables with an easy to use strategy that we did in seconds. I will always remember it.

- Aaron, aged 11

INTRODUCTION

Working hard is <u>not</u> the answer – work smart first, then work hard

When you read this book, I like you to imagine I am sitting next to you talking to you as your personal tutor teaching and sharing these techniques with you.

"The obvious is not obvious until it is obvious"

Does that make sense; Yes or No?

Let me explain what I mean with a few examples.

What is 3 x 4?   3 x 4 equals 12   or   12 = 3 x 4

That is: 1 ….2….3…..4

Should I say that again, it is 1….2….3….4….

(Obvious –isn't it ?)

What is 7 x 8?   It is 56 - So 7 x 8 is 56…and 56 = 7 x 8

So it is 5…6…7…8! ………… So it is 5…6…7…8

Can you see the obvious **IS** now obvious?

1

Not all times tables follow a pattern, so sometimes we may have to use another technique to help us.

We all remember silly things, so say this after me:

3 x 7 is 21 because mum and dad always look 21

Just for fun work tonight, you have to say this 100 times before falling asleep (Joke!)

Here is another one just for fun, say aloud:

8 times 8 is 64 because I ate (8) and ate (8) and

I was Sick (6) on the floor (4),

So 8 eights is 64   That is   8 x 8 = 64

Now say this out loud three times

Louder please.

"*I ate and ate and was sick on the floor, so eight eights is 64*"

You can make up some yourself with a table that you find difficult to remember. Now go and have some fun with it!

We remember silly things, so you have my permission to be 'silly' with your times tables.

HOW DO WE LEARN?

There are 4 stages of learning:-

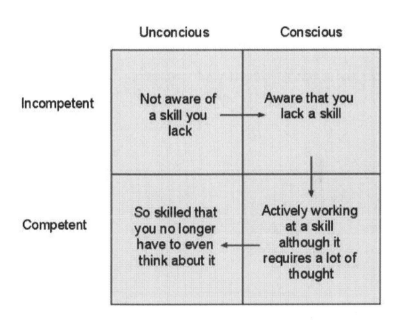

NOTES:-

Fuller Explanation

Unconscious Incompetence:

This is the stage WHEN YOU DON'T KNOW WHAT YOU DON'T KNOW
For this reason many people use a method blindly not knowing if it will work or not and then hope for the best !

Conscious Incompetence:

This is the 'MAKE OR BREAK' Stage.

The individual now has some knowledge, but know they are not good at doing the task. Disappointments can stop their progress. If the individual does not get passed this stage, then their learning is over. For this reason having someone to help them through this stage is essential. By the fact that you are reading this book, you now have ME as your coach to help you.

 I also run live workshops where I can show you how to do it even better whilst having lots of fun!

You can register your interest on

www.the10-secondsmathsexpert.com

Conscious Competence:

This is the 'I CAN DO THIS' Stage.

The individual now understands or knows how to do something. However, demonstrating the skill or knowledge requires concentration and yet they still have to think about each step. It is rather like when you first learn to drive a car or when you first learn how to use a computer.

Unconscious Competence:

This is the 'I AM GREAT' Stage.

The individual has had so much practice with a skill that it has become "automatic" and can be performed easily. They are so good at this level of development that they may even be able to teach others how to do it. It's a bit like how you don't need to think about breathing as for most people it is so automatic.

HOW WE GATHER INFORMATION

Before we get into the techniques, I WANT YOU to know how YOUR brain works, then YOU can do smart things and be more efficient. Does that make sense?

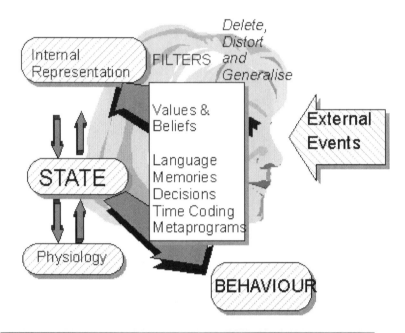

Reticular Activating System (RAS)

Our mind is constantly bombarded by a millions of pieces information received by our 5 senses (visual, auditory, taste, smell, and touch). Luckily, we are equipped with a Reticular Activating System (or RAS) that filters this sensory information. Without our RAS, we would simply be overwhelmed by information overload.

The RAS works by filtering all sensory information we receive, and channelling them into either our unconscious or conscious mind. And this affects the way we learn.

Our RAS will normally do one of three things:-

DELETE - DISTORT - GENERALISE

In this book, I want to help you sharpen your RAS capacity and make your learning easier so that you have more fun.

Notes for reader on 'The 4 Stages of Learning'

To increase your RAS radar, for the next 7 days you might like to do these.

Suggested Exercises

- Pick a colour and notice when you are doing your normal routine, how many objects or people are displaying that colour ;

- Pick a name of a person you are not familiar and notice in your travel how many times THAT NAME shows up ;

- Decide for yourself to focus on something or see how many things you started to noticed but in the past you have walked by without paying any attention to it!

Have fun!

ENRICHMENT FOR BASIC MATHS

(For older pupils, Chapters 1 to 3 is for additional information)

CHAPTER ONE

Train your brain spotting patterns in times tables to increase your memory power

Patterns in Times Tables (Part One)

Train your brain to recognise patterns in times tables to increase memory power - creating a mental 'hook'

0 and 10 times tables

1 and 9 times tables

2 and 8 times tables

4 and 6 times tables

3 and 7 times tables

Task one for Your RAS

What do you notice about the pairing of the times tables in the above list:-

0 and 10, 1 and 9, 2 and 8, 4 and 6, 3 and 7?

Answer: Each paring of numbers add up to 10. Well done if you spotted the pattern.

The obvious is not obvious, until it's obvious - Right?

Let's explore this further and look at times table in a slightly different way.

Starting with the easy ones first!

0 and 10 Times table

1 x 0 = 0	1 x 10 = 10
2 x 0 = 0	2 x 10 = 20
3 x 0 = 0	3 x 10 = 30
4 x 0 = 0	4 x 10 = 40
5 x 0 = 0	5 x 10 = 50
6 x 0 = 0	6 x 10 = 60
7 x 0 = 0	7 x 10 = 70
8 x 0 = 0	8 x 10 = 80
9 x 0 = 0	9 x 10 = 90
10 x 0 = 0	10 x 10 = 100

Task two for your RAS

What did you notice?

If you noticed that the last digit ends in '0' – well done again!

You are beginning to understand that "The obvious is not obvious until it's obvious!"

Most importantly you are exercising your RAS which marks you out as a genius compared to other ordinary people!

Say out aloud: "I AM A GENIUS. Yep!"

In fact, do this every day and as often as possible because

"YOU ARE AMAZING!"

<u>1 and 9 Times table</u>

Look at the last digit (unit position)

$1 \times 1 = 01$ 1

$2 \times 1 = 02$ 2

$3 \times 1 = 03$ 3

$4 \times 1 = 04$ 4

$5 \times 1 = 05$ 5

$6 \times 1 = 06$ 6

$7 \times 1 = 07$ 7

$8 \times 1 = 08$ 8

$9 \times 1 = 09$ 9

$10 \times 1 = 10$ 1

(This is when you keep adding the numbers until you get a single digit.)

	Casting out numbers
$1 \times 9 = 09$	$0 + 9 = 9$
$2 \times 9 = 18$	$1 + 8 = 9$
$3 \times 9 = 27$	$2 + 7 = ?$
$4 \times 9 = 36$	$3 + 6 = ?$
$5 \times 9 = 45$	$4 + 5 = ?$
$6 \times 9 = 54$	$5 + 4 = ?$
$7 \times 9 = 63$	$6 + 3 = ?$
$8 \times 9 = 72$	$7 + 2 = ?$
$9 \times 9 = 81$	$8 + 1 = ?$
$10 \times 9 = 90$	$9 + 0 = ?$

Task 3 for Your RAS

This time, did you notice several things from the 1 and 9 times tables?

The 9 times tables got lots of interesting patterns; take your time and look for it! Do some research and you will be fascinated by what you will find.

Discovery

I hope you found at least 3 things in the 9 times table:

1. The ascending (going up) of the TENs digits and descending (going down) of the digits in the UNITs.

2. The answers of the 9 times tables add up to 9- this is done by 'Casting out'
   (0+9, 1+8, 2+7, 3+6, 4+5, 5+4, 6+3, 7+2, 8+1, 9+0 all add up to 9)

3. The 'Mirror' pattern :
09	90
18	81
27	72
36	63
45	54

"Give a man a fish and he will eat for a day – teach him to fish and he will eat for a LIFETIME!"

When you learn to discover things for yourself, this is the best form of learning and another great way to learn is to share it with someone…..a
friend…..mum….dad….brothers…sisters…your teacher.

You can email me to share any fun things you notice, in return I will show you something special to say 'Thank you' just for you.

info@the10-secondsmathsexpert.com

Discovery Task for you

Investigate the patterns for yourself:-

a) 2 and 8 times tables
b) 4 and 6 times tables
c) 3 and 7 times tables
d) 2, 4 and 6 times tables

Have fun!

Discovery Task results

Let me share with you my findings, but if you have found other patterns then well done and please share it with me and go to: www.the10-secondsmathsexpert.com

2 and 8 times tables (Look at the last UNIT digits)

1 x 2 = 02	2			
2 x 2 = 04		4		
3 x 2 = 06			6	
4 x 2 = 08				8
5 x 2 = 10					0
6 x 2 = 12	2			
7 x 2 = 14		4		
8 x 2 = 16			6	
9 x 2 = 18				8
10 x 2 = 20					0
1 x 8 = 08	8			
2 x 8 = 16		6		
3 x 8 = 24			4	
4 x 8 = 32				2
5 x 8 = 40					0
6 x 8 = 48	8			
7 x 8 = 56		6		
8 x 8 = 64			4	
9 x 8 = 72				2
10 x 8 = 80					0

4 and 6 times tables (Again look at the last UNIT digits)

1 x 4 = 04	4	1 x 6 = 06	6
2 x 4 = 08	8	2 x 6 = 12	2
3 x 4 = 12	2	3 x 6 = 18	8
4 x 4 = 16	6	4 x 6 = 24	4
5 x 4 = 20	0	5 x 6 = 30	0
6 x 4 = 24	4	6 x 6 = 36	6
7 x 4 = 28	8	7 x 6 = 42	2
8 x 4 = 32	2	8 x 6 = 48	8
9 x 4 = 36	6	9 x 6 = 54	4
10 x 4 = 40	0	0 x 6 = 60	0

Can you put these results in a simpler way to make it easier to remember?

I deliberately did not put it in the 'best' way because I want to challenge you to come up with your own way to represent the result. Research it.

Alternatively when you get a chance to come to my live workshops, I can share with you others ways to make it more interesting. To register your interest in attending one of my workshops, you can send me an email to:

info@the10-secondsmathsexpert.com

Reader's notes on Patterns for 4 and 6 Times Tables

3 and 7 times tables (These two tables have different patterns).Discover it for yourself before I reveal it to you on the next page.

$$1 \times 3 = 03$$
$$2 \times 3 = 06$$
$$3 \times 3 = 09$$
$$4 \times 3 = 12$$
$$5 \times 3 = 15$$
$$6 \times 3 = 18$$
$$7 \times 3 = 21$$
$$8 \times 3 = 24$$
$$9 \times 3 = 27$$
$$10 \times 3 = 30$$

$$1 \times 7 = 07$$
$$2 \times 7 = 14$$
$$3 \times 7 = 21$$
$$4 \times 7 = 28$$
$$5 \times 7 = 35$$
$$6 \times 7 = 42$$
$$7 \times 7 = 49$$
$$8 \times 7 = 56$$
$$9 \times 7 = 63$$
$$10 \times 7 = 70$$

Now, what did I say?

"The obvious is not obvious until It's obvious!"

PERSONAL NOTES

There is another pattern when you look at the part of the 7 and 3 times tables side by side

1 x 7 = 07	9 x 3 = 27	
2 x 7 = 14	8 x 3 = 24	
3 x 7 = 21	7 x 3 = 21	
4 x 7 = 28	6 x 3 = 18	7 – 4 – 1 – 8
		PATTERN

Did you see the pattern, the MAIN PATTERN of the 3 and 7 times tables?

If not, then let's RAS up your RAS!

1 x 3 = 03

2 x 3 = 06   that's right, it increases by 3

3 x 3 = 09

4 x 3 = 12

5 x 3 = 15   this also increases by 3

6 x 3 = 18

7 x 3 = 21

8 x 3 = 24   and again, it increases by 3 in groups of 3's

9 x 3 = 27

10 x 3 = 30

1 x 7 = 07

2 x 7 = 14   same thing, it increases by 3 in groups

3 x 7 = 21

4 x 7 = 28

5 x 7 = 35

6 x 7 = 42

7 x 7 = 49

8 x 7 = 56

9 x 7 = 63

10 x 7 = 70

Was there something else you noticed about the answers in the 3 and 7 times tables? Use you RAS…RAS…RAS

NOTES

PERSONAL NOTES

CHAPTER TWO

Train your brain spotting further patterns in times tables

Patterns in Times Tables (Part Two)

2,4,6,8 times tables

5 and 10 times tables

PATTERNS FOR 2, 4, 6 and 8 times tables

(There is an old saying: 2...4...6...8...who do we appreciate?

....mum....dad...your best friend.....me?)

Task 4 for Your RAS (2, 4, 6 and 8 times tables)

To help you, I will show you the tables, and then it is your turn to tell me what do you notice? KISSY...KISSY please........no, I don't mean that!

K(eep) I(t) S(imple) S(sweetheart) ....Y(ou brilliant person !)

1 x 2 = 2	1 x 4 = 4
2 x 2 = 4	2 x 4 = 8
3 x 2 = 6	3 x 4 = 12
4 x 2 = 8	4 x 4 = 16
5 x 2 = 10	5 x 4 = 20
6 x 2 = 12	6 x 4 = 24
7 x 2 = 14	7 x 4 = 28
8 x 2 = 16	8 x 4 = 32
9 x 2 = 18	9 x 4 = 36
10 x 2 = 20	10 x 4 = 40

2...4...6...8...who do we appreciate?

1 x 6 = 6	1 x 8 = 8
2 x 6 = 12	2 x 8 = 16
3 x 6 = 18	3 x 8 = 24
4 x 6 = 24	4 x 8 = 32
5 x 6 = 30	5 x 8 = 40
6 x 6 = 36	6 x 8 = 48
7 x 6 = 42	7 x 8 = 56
8 x 6 = 48	8 x 8 = 64
9 x 6 = 54	9 x 8 = 72
10 x 6 = 60	10 x 8 = 80

2...4...6...8...who do we appreciate?

Did you see the clues I gave you?

If you have worked it out yourself – Well done!

You are now really RAS-SING it up!

5 and 10 times tables

By now, your RAS is so sharp, I hope you will see the 'obvious' and can tell me the 'obvious' in the 5 times table and the 'obvious' in the 10 times table.

In fact, better still, go and share what you have noticed with a member of your family or a friend or even your teacher.

1 x 5 = 05	1 x 10 = 10
2 x 5 = 10	2 x 10 = 20
3 x 5 = 15	3 x 10 = 30
4 x 5 = 20	4 x 10 = 40
5 x 5 = 25	5 x 10 = 50
6 x 5 = 30	6 x 10 = 60
7 x 5 = 35	7 x 10 = 70
8 x 5 = 40	8 x 10 = 80
9 x 5 = 45	9 x 10 = 90
10 x 5 = 50	10 x 10 = 100

To obtain your Bonus, go to:

www.the10-secondsmathsexpert.com

Bonus 1: Special Dice Game for the family

Pick Square numbers !

Explore other Patterns in Times Tables, using the ONE DOWN – ONE UP Investigation

DECREASE First digit by one and INCREASE the Second digit by one

For example :
6 x 6 = 36
5 x 7 = 35

8 x 8 = 64
7 x 9 = 63

Does this always work ?

Investigate further.

What happens if you DECREASE First digit by two and INCREASE the Second digit by two ?
What do you notice ?
Test it out and have fun investigating the DOWN & UP!

PERSONAL NOTES

PERSONAL NOTES

CHAPTER THREE

Patterns in 'CASTING OUT' – Know you are Right!

"One sign of a genius is when you SEE what ordinary people don't see and when you show them the obvious, they will say to you: 'You're a genius'

Let that be <u>YOU</u> when you finish learning from this book.

We are now going to start RAS-SING it up even further using TSBD (The Same But Different)

Let's look at these times tables using CASTING OUT method.

3 and 6 times tables
4 times table
8 times table
9 times table

What is 'Casting Out'?

Casting out nines
From Wikipedia, the free encyclopaedia

Casting out nines is a ***sanity test*** to ensure that hand computations of sums, differences, products, and quotients of ***integers*** are correct. By looking at the ***digital roots*** of the inputs and outputs, the casting-out-nines method can help one check arithmetic calculations. The method is so simple that most school children can apply it without understanding its mathematical underpinnings.

The method involves converting each number into its "casting-out-nines" equivalent, and then redoing the arithmetic. The casting-out-nines answer should equal the casting-out-nines version of the original answer. Below are examples for using casting-out-nines to check ***addition, subtraction, multiplication,*** and ***division.***

Did you understand all that mumbo jumbo of a more formal definition of 'casting out'?

So let's RAS-it to make simple sense of it.

Let's do a few examples to help us understand how to 'Cast out' any numbers.

For example:

12 ....1 + 2 = 3

so when we cast out 12, the final single digit is 3

234........... 2 + 3 + 4 = 9

2356..........2 + 3 + 5 + 6 = 16......carry on...... and 16 .....1 + 6 = 7

Try these for yourself (you should always end on a single digit)

a) 35 b) 256 c) 2574   d) 23467   e) 537894

Answer :

a) 8 b) 4 c) 9 d) 4 e) 9

Patterns in Casting out
3 and 6 times tables

$1 \times 3 = 3$.. 3

$2 \times 3 = 6$... 6

$3 \times 3 = 9$... 9

$4 \times 3 = 12$...... $1 + 2$... 3

$5 \times 3 = 15$...... $1 + 5$.. 6

$6 \times 3 = 18$...... $1 + 8$.. 9

$7 \times 3 = 21$....... $2 + 1$.................................... 3

$8 \times 3 = 24$....... $2 + 4$.................................... 6

$9 \times 3 = 27$....... $2 + 7$.................................... 9

$10 \times 3 = 30$.....$3 + 0$... 3

$11 \times 3 = 33$.....$3 + 3$... 6

$12 \times 3 = 36$.....$3 + 6$... 9

$13 \times 3 = 39$.....$3 + 9 = 12$.......$1 + 2$................... 3

$14 \times 3 = 42$.......$4 + 2$................................ 6

$15 \times 3 = 45$.....$4 + 5$... 9

etc...

1 x 6 = 6..   6

2 x 6 = 12..   3

3 x 6 = 18..   9

4 x 6 = 24.......2 + 4.................................   6

5 x 6 = 30...... 3 + 0.................................   3

6 x 6 = 36.......3 + 6.................................   9

7 x 6 = 42.......4 + 2.................................   6

8 x 6 = 48.......4 + 8 = 12 ....1 + 2   3

9 x 6 = 54.......5 + 4.................................   9

10 x 6 = 60......6 + 0.................................   6

11 x 6 = 66......6 + 6 = 12........1 + 2................   3

12 x 6 = 72......7 + 2.................................   9

13 x 6 = 78......7 + 8 = 15.........1 + 5............   6

14 x 6 = 84......8 + 4 = 12...... 1 + 2.................   3

15 x 6 = 90......9 + 0..................................   9

etc...

Task 5 for Your RAS

Discover about the patterns for the 4 and 8 times tables when you are 'Casting out' the answer.. Hint: The 4 times table is a bit more tricky to see the pattern.

Patterns for Task 5 RAS

I hope you have seen the pattern. See if you agree with me or did you find something else. If so, share it with me on info@the10-secondsmathsexpert.com

Let's look at the easier one, that is the 8 Times Table.

	Casting out	Final Pattern
1 x 8 = 08	8	8
2 x 8 = 16	7	7
3 x 8 = 24	6	6
4 x 8 = 32	5	5
5 x 8 = 40	4	4
6 x 8 = 48	12 (1 + 2)	3
7 x 8 = 56	11 (1 + 1)	2
8 x 8 = 64	10 (1 + 0)	1
9 x 8 = 72	9	9
10 x 8 = 80	8	8
11 x 8 = 88	16 (1 + 6)	7
12 x 8 = 96	15 (1 + 5)	6
13 x 8 =104	5	5
14 x 8 =112	4	4

Now you continue and complete the table all the way
 to 20 x 8

15 x 8 =

16 x 8 =

17 x 8 =

18 x 8 =

19 x 8 =

20 x 8 =

If you extend the table, would the pattern continue?
Investigate.

Now for the interesting 4 times table and the pattern

Casting out

1 x 4 = 4	4
2 x 4 = 8	8
3 x 4 = 12	3 (1+2)
4 x 4 = 16	7 (1+6)
5 x 4 = 20	2 (2+0)
6 x 4 = 24	6 (2+4)
7 x 4 = 28	1 (2+8=10…..1)
8 x 4 = 32	5 (3 +2)
9 x 4 = 36	9 (3+6)
10 x 4 = 40	4 (4+0)
11 x 4 = 44	8 (4+4)
12 x 4 = 48	3 (4+8=12……3)

Do you see the pattern?

"The obvious is not obvious until it's obvious!"

Let's learn to look at this another way!

The pattern hidden within a pattern

1 x 4 = 4	4......................4
2 x 4 = 8	8..8
3 x 4 = 12	3......................3
4 x 4 = 16	7..7
5 x 4 = 20	2......................2
6 x 4 = 24	6..6
7 x 4 = 28	1......................1
8 x 4 = 32	5..5
9 x 4 = 36	9 9
10 x 4 = 40	4..4
11 x 4 = 44	8......................8
12 x 4 = 48	3..3

You can continue this and see what happens. Try it!

ENRICHMENT FOR QUICK CALCULATION

This section will help you with your examination techniques

Like KS2 SAT Maths examinations if you are in schools in the UK.

CHAPTER FOUR

Times tables in your hands (PART ONE)

Times tables at your hand: "Your handy tools !"

Your 'Hand Calculator'

For 9 times tables only

A simple way to remember your 9 times table

Hand Calculator for 9 times tables

Instruction:

Spread your fingers as shown by the diagram

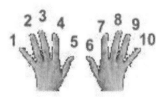

Count from left to right and as you do so, bend the fingers.

Count : 1..2..3..4..5.(left hand) and continue with (right hand)

6..7..8..9..10

Do this once more

Count : 1..2..3..4..5.(left hand) and continue with (right hand) 6..7..8..9..10

Let's do an example to see how it works

To work out, say for example: 3 x 9

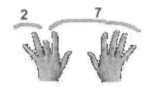

You bend the third finger as shown. There are 2 fingers to the LEFT of the bent finger and 7 fingers altogether to the RIGHT of the bent finger.

So 3 x 9 = 2 (left) 7 (right) = 27

That's it!

Let's do another one, say 6 x 9

From the starting position:

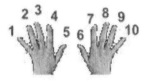

Now count from left to right 1...2...3...4...5...6 (bend the 6th finger)

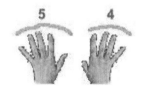

Ask yourself, how fingers to the left of the bent finger (5) and how many to the right (4)

So 6 x 9 = 5 (left) 4(right) = 54

Check it out with the NINE TIMES TABLE

<u>Task 6 for Your RAS Practice</u>

Bent Finger		Left of Bent Finger	Right of Bent Finger
1	x 9	= 0	9
2	x 9	= 1	8
3	x 9	= 2	7
4	x 9	= 3	6
5	x 9	= 4	5
6	x 9	= 5	4
7	x 9	= 6	3
8	x 9	= 7	2
9	x 9	= 8	1
10	x 9	= 9	0

PERSONAL NOTES

PERSONAL NOTES

CHAPTER FIVE

Times tables in your hands (PART TWO)

For 6 , 7, 8 and 9 times tables

A simple way to remember your 6, 7, 8 and 9 times table all at once!

Finger Multiplications for 6...7...8...9 Times Tables (all at once)....WOW ! I hope you will say!

Here is a great way to do all the tables above in one go.

It will look complicated but once you have done it a few times slowly, you will be amazed how easy it will be. Let's begin................

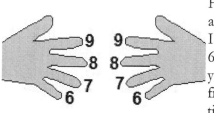

Hold your hands like this and
I want you to count 6,7,8,9,10 (thumbs). As you count, bend the fingers at exactly the same time.
Do this 3 times slowly.

For Example
If you want to do 8(left) x 7(right), then let the finger number 8 touch the finger number 7 on the other hand. Now count think TOUCHING FINGERS AND BELOW and fingers ABOVE the touching fingers. Count each of these as a 10

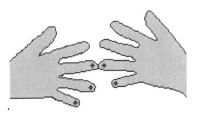

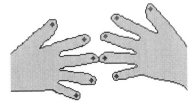

To get the answer. First you add the fingers touching and below , altogether there are 5 fingers, so that makes 5x10 = 50. That is 5 fingers each representing 10, so 5 x 10=50

For the fingers above the touching finger, in the LEFT hand are 2 fingers and times this to the

RIGHT hand 3 fingers, so 2 x 3 = 6

Task 7 for Your RAS Practice

I would suggest you get use to this technique by doing the exercise in this order:

7 x 7   8 x 8   9 x 9   and then you can do it in any order you want.

For example: 8 x 6   6 x 9   7 x 8   BUT leave 6 x 6 and 6 x 7 to last!

---

6 x 6 :

When you touch 6(on left hand) and 6 (on right hand) together, there are 2 fingers touching, so 2 Tens equals 20. Now there are 4 fingers above on the Left and Right hands, so 4 times 4 equals 16.

So the final answer is 20 + 16 = 36   So 6 x 6 = 36

One of the key to success is repetitive practice –

Pick one or two Times Tables

and practice them for ONE MINUTE EACH DAY

(No more) and repeat the next day !

6 x 1 =
6 x 2 =
6 x 3 =
6 x 4 =
6 x 5 =
6 x 6 =
6 x 7 =
6 x 8 =
6 x 9 =
6 x 10 =
6 x 11 =
6 x 12 =

8 Times Table Practice

8 x 1 =
8 x 2 =
8 x 3 =
8 x 4 =
8 x 5 =
8 x 6 =
8 x 7 =
8 x 8 =
8 x 9 =
8 x 10 =
8 x 11 =
8 x 12 =

PERSONAL NOTES

CHAPTER SIX

New ways to do 3, 6, 7 and 9 times tables

FUN TECHNIQUES

In this chapter, we have 2 fun techniques.

1) The Snake and Ladder method for 7 times table in under 10 seconds!

2) How to do learn your 3, 6 and 9 times tables all at once in under 10 seconds!

WOW !!

SNAKE AND LADDER METHOD FOR THE 7 TIMES TABLE

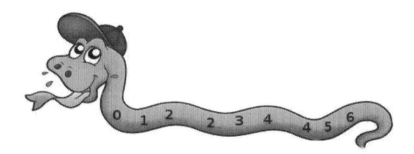

THE SNAKE (in 3 PARTS):

STEP 1 :PUT IN NUMBERS ON THE SNAKE

Like this:

0	1	2
2	3	4
4	5	6

STEP 2 : PUT IN LADDER NUMBERS

Like this

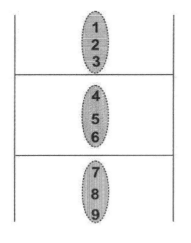

7	4	1
8	5	2
9	6	3

THE COMPLETE DIAGRAM

(07)	14	21
28	35	42
49	56	63
		(70)

SNAKE AND LADDER METHOD FOR THE 7 TIMES TABLE (Practice Sheet)

THE SNAKE (in 3 PARTS) - Now it's YOUR TURN!

STEP 1: PUT IN NUMBERS ON THE SNAKE

STEP 2: PUT IN LADDER NUMBERS

THE COMPLETE
DIAGRAM

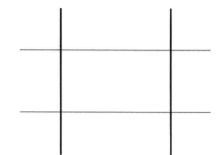

Task 8 for Your RAS
Practice 'Snake &
Ladder'

PRACTICE SPACE

7 Times Table Practice

7 x 1 =
7 x 2 =
7 x 3 =
7 x 4 =
7 x 5 =
7 x 6 =
7 x 7 =
7 x 8 =
7 x 9 =
7 x 10 =
7 x 11 =
7 x 12 =

How to do your 3-6-9 Times Tables all at once in under 10 seconds!

Instruction :

1) Start by drawing 3 grids and insert the numbers like this :

3	6	9
2	5	8
1	4	7

3	6	9
2	5	8
1	4	7

3	6	9
2	5	8
1	4	7

2) Now this is the clever bit!

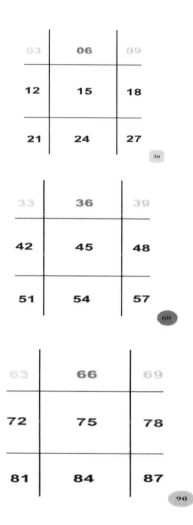

So, here is your 3 times table up to 3 x 30 = 90!

Can you see where the 6 times table is?

Look carefully!

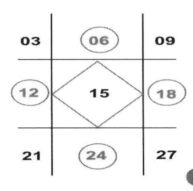

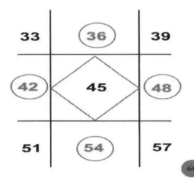

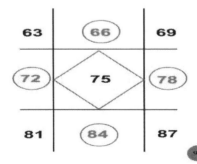

What about the 9 times table?

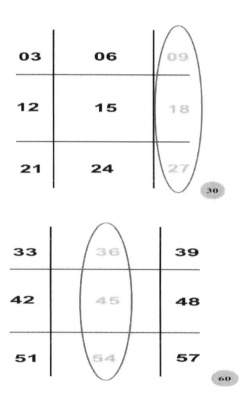

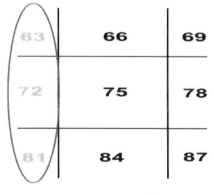

PERSONAL NOTES

"The obvious is not obvious until

it's obvious !"

Task 9 for Your RAS Practice

Now it's your turn.
What does Practice make?
No Practice makes it WORSE!
Only Perfect Practice makes it Perfect!

3	6	9
2	5	8
1	4	7

3	6	9
2	5	8
1	4	7

3	6	9
2	5	8
1	4	7

And do it once more:

3	6	9
2	5	8
1	4	7

3	6	9
2	5	8
1	4	7

3	6	9
2	5	8
1	4	7

Do this ONE MINUTE per day for the next 7 days and you will know your 3-6-9 times tables.

Task 9 for Your RAS Practice   (Practice Sheet)

Now it's your turn.

What does Practice make?

No Practice makes it WORSE!

Only Perfect Practice makes it Perfect!

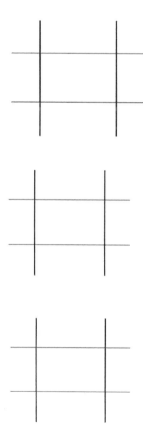

And do it once more:

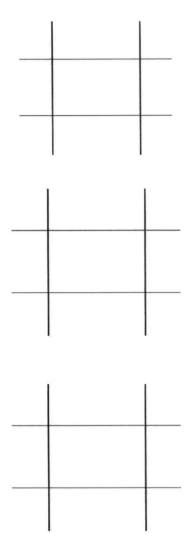

Do this ONE MINUTE per day for the next 7 days and you will know your 3-6-9 times tables.

CHAPTER SEVEN
The power of doubling

- The power of doubling
- The simple way to learn your 12 times table
- Multiply any number by 12 in seconds

In this chapter you will see how to apply the 2 times table (or doubling) in a very powerful way!

To warm up your NOODLES, that's the term I use for our Brain!

Let's do a very simple exercise by doubling the numbers below:

Number	Double the number	Number	Double the number	Number	Double the number
3		24		112	
14		21		213	
23		52		431	
41		72		1243	

The key exercise (If you can do this 12 times table and multiplying by 12 it is easy!)

So just double the numbers and put the answer underneath in the box below:

1	2	3	4	5	6	7	8	9	10

11	12	13	14	15	16	17	18	19	20

12 times table	What is double	Answer
1 x 12	1 (double 1 is 2)	12
2 x 12	2 (double 2 is 4)	24
3 x 12	3 (double 3 is 6)	36
4 x 12	4 (double 4 is 8)	48
5 x 12	5(double 5 is 10, so add 1 to 5 makes 6)	60
6 x 12	6(double 6 is 12, so add 1 to 6 makes 7)	72
7 x 12	7(double 7 is 14, so add 1 to 7 makes 8)	84
8 x 12	8(double 8 is 16, so add 1 to 8 makes 9)	96
9 x 12	9(double 9 is 18, so add 1 to 9 makes 10)	108
10 x 12	10(double 10 is 20, so add 2 to 10 makes 12)	120
11 x 12	11(double 11 is 22, so add 2 to 11 makes 13)	132
12 x 12	12(double 12 is 24, so add 2 to 12 makes 14)	144

Quick 12 -How to multiply any number by 12 quickly

Key skills:

1) To be able to double the number you want by 12
2) To be able to add

Question	Double this number	Write it like this	Put down the last digit at the end and add together the numbers underlined to get the final answer
14 x 12	14	_14_ / 2_8_ (14+2 = 16)	168
21 x 12	21	_21_ / _4_2 (21+4=25)	**252**
34 x 12	34	_34_ / _1_2 (34 + 1=35)	**352**
53 x 12	53	_53_ / _10_6 (53 +10 = 63)	**63 6**
67 x 12	67	_67_ / _13_4 (67 + 13 = 80)	**804**
123 x 12	123	_123_ / _24_6 (123+24 =147)	**1476**

Now it is your turn to have ago at these questions (you can check it with a calculator after you have done it on paper :

1) 24 x 12   2) 35 x 12   3) 47 x 12   4) 72 x 12

5) 114 x 12

Remember   24 x 12 is the same as 12 x 24, you double the number that is multiplied by 12.

To obtain your Bonus, go to:

www.the10-secondsmathsexpert.com

Bonus 2: How to multiply without multiplying – using straight lines only.

ENRICHMENT FOR BASIC EXAM SUCCESS

CHAPTER EIGHT

Special Subtraction method –Take away in a flash!

Take away without taking away!

A simple quick method to subtract from 100, 1 000, 10 000, 100 000.........etc

An ancient method to take away in seconds

QUESTION :

If I asked you, can you do this sum in your head or on paper within 5 seconds

$$1\ 000\ 000\ 000\ 000\ -\ 835\ 364\ 764\ 362\ ?$$

For many people, the answer is ' No way '. ...or something to that effect!"

In this chapter, I will show you how to do this with only a few minutes practise and you can amaze your friends!

Shall we get started?

Here is a simple "rule" to remember:

ALL TO 9
LAST TO 10
(Make only the last number to 10)

---

and I would like you to say this aloud 3 times.

Do it NOW!

Let's warm you up to do this exercise.

Write down below what numbers you must add to make up to 9:

Number	0	1	2	3	4	5	6	7	8	9
Add to make 9										

Next, write down below what numbers you must add up to make up to 10:

Number	0	1	2	3	4	5	6	7	8	9
Add to make 10										

How does it work?

Here are a few examples:

To help you, remember:

1. Only add to the last digit (the 'unit' position to make up to 10)
2. All other numbers you need to add to make 9.

100 -   24   = 76   (2+7 =9   4+6 =10)

1000 -   527   =473   (5+4=9 2+7=9   7+3=10)

10000 -   3826= 6174 (3+6=9 8+1=9   2+7=9 6+4=10)

Or you can write it this way:

1 00	1 000	10 000
- 24	- 527	- 3826
76	473	6174

To help you with the skill, I have broken it down into a few mini exercises:

1	0	0	-	4	8 =		
1	0	0	-	7	6 =		
1	0	0	-	8	3 =		

1	0	0	0	-	3	5	7 =		
1	0	0	0	-	7	2	8 =		
1	0	0	0	-	8	0	6 =		

Now you have the key ideas, to improve your speed and to amaze your friends and family.

PRACTICE-PRACTICE-PRACTICE

Remember :

Practice (incorrect techniques) only makes it WORSE.
However
"PERFECT PRACTICE MAKES PROGRESS PERFECT! "

Do this exercise and use a calculator to check the answers, then you can impress anyone:

1.   100 - 58

2.   1,000 - 528

3.   10,000 - 6248

4.   100,000 - 67254

5.   100,000 - 32577

6.   1 000 000 - 243 563

7.   1 000 000 - 543 257

8.   10 000 000 - 5 356 388

9.   100 000 000 - 47 814 638

10. 100 000 000 000 - 57 247 816 2 88

How did you get on?

Did you find that you actually took longer to use a calculator to check the answers when doing this exercise? If that is the case...

WOW! You are a genius!

CHAPTER NINE

Quick mental multiplying technique –Double your mind power.

A) Three step method for multiplying numbers between : 10 - 20 and 20 - 30.

B) Using the 3 step method to work out all the square numbers less than 30.

A) For multiplying numbers between 10 - 20
B) Let's use an example, say 14 x 15

Step 1 -Add the second digit number (5) to the first number (14) : 14 + 5 =19

Step 2 – From the answer of Step 1 (19),

multiply by 10

$$19 \times 10 = 190$$

Step 3- Multiply the ' digits '(underlined) of both numbers 4 x 5 = 20,

then add it to 190 ,   190 + 20 = 210

So   14 x 15 = 210

It may look long, but once you have done a few examples, you will be amazed how quickly you can do this.

Here is another example, to make sure you know the technique

13 x 18

Step 1 - 13 + 8 = 21

Step 2 - 21 x 10 = 210

Step 3 - 3 x 8 = 24, then add it to 210
...........210 + 24 = 234

Now it is your turn.

Do it the long way first and after a few times you will find you should be able to do it mentally.

Do this exercise and then check the answers using a calculator.

$$12 \times 15$$

$$13 \times 17$$

$$16 \times 18$$

$$14 \times 19$$

$$18 \times 19$$

Using the Three Step Method to work out Square Numbers from 11 to 19

	Square Number	Add the number to Unit digit (underlined)	Multiply last answer by 10	Square the unit digit	Final Answer
11 x 11	11^2	11 + 1 = 12	12 0	$1^2 = 1$	120 + 1 = 121
12 x 12	12^2	12 + 2 = 14	140	$2^2 = 4$	140 + 4 = 144
13 x 13	13^2	13 + 3 = 16	160	$3^2 = 9$	160 + 9 = 169
14 x 14	14^2	14 + 4 = 18	180	$4^2 = 16$	180 + 1(+196
15 x 15	15^2				
16 x 16	16^2				
17 x 17	17^2				
18 x 18	18^2				
19 x 19	19^2				

C) For multiplying numbers between 20 -30
   Let's use an example, say 2<u>4</u> x 2<u>5</u>

Step 1 -Add the second digit number (5) to the
first number (14) : 24 + 5 =2 9

Step 2 – From the answer of Step 1 (29), this
time multiply by 20

29 x 20 = 580

Step 3- Multiply the 'digits' (underlined) of both
numbers 4 x 5 = 20,

then add it to 190,   580 + 20 = 600

So   24 x 25 = 600

Here is another example, to make sure you
know the technique

23   x   28

Step 1 -   23 + 8 = 31

Step 2 -   31 x 20 = 620

Step 3 -   3 x 8 = 24, then add it to 620
..........620 + 24 = 644

Do this exercise and then check the answers using a calculator.

1. 22 x 25
2. 23 x 27
3. 26 x 28
4. 24 x 29
5. 28 x 29

Using the Three Step method to work out Square Numbers from 21 to 29

	Square Number	Add the number to Unit digit (underlined)	Multiply the last answer by 20	Square the unit digit	Final Answer
21 x 21	21^2	21 + 1 = 22	440	$1^2 = 1$	440 + 1 = 441
22 x 22	22^2	22 + 2 = 44	480	$2^2 = 4$	480 + 4 = 484
23 x 23	23^2	23 + 3 = 26	520	$3^2 = 9$	529
24 x 24	24^2	24 + 4 = 28	560	$4^2 = 16$	560 + 16 = 576
25 x 25	25^2				
26 x 26	26^2				
27 x 27	27^2				
28 x 28	28^2				
29 x 29	29^2				

PERSONAL NOTES

CHAPTER TEN

Do FRACTIONS in a fraction of the time!

Using the 'Battenburg' technique for adding fractions is a piece of cake!

For many people, dealing with FRACTIONS in maths is a complicated task.

However, in this chapter, I will show you how to ADD FRACTIONS using the Battenburg techniques which will take seconds.

Let see how it works:

Take an Example : $\underline{2}$ + $\underline{3}$

5   11

Now draw a 2 by 2 grid like this:

	3	11
2	x	22
5	15	55

Now add the numbers on the diagonal together:

15 + 22 = 37

Therefore the answer is 37 / 55

FUN MATHS FOR ALL AGES

<u>Special 4 by 4 Birthday Magic square</u>

The 4 by 4 Magic square has got some very interesting features

This is a JB Special (J-ust B-rilliant Special) !

Suppose a member of your family was born for example on 6th May 1959.

We can use this to form a 4 by 4 Magic square.

Rewrite date of birth as: 06 -05- 1959

Use these numbers to form the Magic square like this

To work out the Magic Total = 6 + 5 + 19 + 59 =89

Therefore, each row, each column and diagonals should add up to 89.

Please check it with or without a calculator.

6	5	19	59
20	58	7	4
57	17	7	8
6	9	56	18

Challenge:

How many different 4 numbers in this magic square add up to 89?

For Example: 6 + 59 + 6 + 18 = 89

6	5	19	59
20	58	7	4
57	17	7	8
6	9	56	18

Have a go ?

To obtain your Bonuses, go to:

www.the10-secondsmathsexpert.com

Bonus 3: The 4 by 4 Magic Square system revealed.

Bonus 4: Special Tribute to Nelson Mandela (His Special 4 by 4 Magic Squares)

Special Bonus Chapter

Let me share with you another idea to learn your 8 times tables.

Here is the normal 8 times table :

$$
\begin{array}{rcl}
8 \times 1 &=& 08 \\
8 \times 2 &=& 16 \\
8 \times 3 &=& 24 \\
8 \times 4 &=& 32 \\
8 \times 5 &=& 40 \\
8 \times 6 &=& 48 \\
8 \times 7 &=& 56 \\
8 \times 8 &=& 64 \\
8 \times 9 &=& 72 \\
8 \times 10 &=& 80
\end{array}
$$

What did you notice about the last column (the UNIT DIGITS)?

If you have seen the last digits are :

8-6-4-2-0 and it repeats 8-6-4-2-0 if you extend the 8 times table. Well done !

This gives us a key how to learn the 8 times table another way !

This is what I want you to do :

Write, in first column : 8 7 6 5 4 4 3 2 1 0
Write, in the Second column : 0 2 4 6 8   0 2 4 6 8

It should look like this :

FIRST COLUMN	SECOND COLUMN
8	0
7	2
6	4
5	6
4	8
4	0
3	2
2	4
1	6
0	8

If you look carefully, this is the 8 times table in reverse :

$$80 = 8 \times 10$$
$$72 = 8 \times 9$$
$$64 = 8 \times 8$$
$$56 = 8 \times 7$$
$$48 = 8 \times 6$$
$$40 = 8 \times 5$$
$$32 = 8 \times 4$$
$$24 = 8 \times 3$$
$$16 = 8 \times 2$$
$$08 = 8 \times 1$$

So by writing in a column :
8 7 6 5 4 4 3 2 1 0

Then second column :

0 2 4 6 8   0 2 4 6 8

You can now do practise your 8 times table quickly !
Enjoy and have fun !

PERSONAL NOTES

PERSONAL NOTES

PERSONAL NOTES

Your mathematical journey – what's next?

Well done for working through this book.

Learning is a never ending journey and you will never stop learning.

The value of learning is about putting it into actions and sharing is fun. As you learn, please share it with others because every time you teach someone else anything you have learnt, you actually reinforced your own learning.

The more you give, the better you will be to be quicker and amaze all your friends.

Many of the techniques I have shared with you in this book are methods I have taught to thousands of people during my teaching career and in hundreds of my maths workshops working with pupils and parents, teachers and other professional colleagues in industry.

Remember that on its own Practice DOES NOT make perfect. It is

PERFECT PRACTICE MAKES PROGRESS THEN PERFECT.

Repetition is the key to improve any skills and the techniques. This book will help you with your mental calculations, as well as performing in mathematics examination.

I hope reading this book has given you an insight to the different ways to learn.

The world of mathematics is a fascinating one and there are many more wonderful discoveries to do the same thing quicker, easier and more fun. Do your own research and you will be amazed at what other wonderful methods there are in the History of Maths. Many great ancient techniques have been forgotten and you might be the person to re-discover a lost technique.

Imagine you could be famous for this discovery – WOW!

I thank you for sharing this journey with me and I hope if you will continue to learn more.

I live in England, UK. If there is sufficient interest, I would love to come to your country to share these and many more 10-Seconds Speed Maths Techniques with you and your friends. Come and join me at some of the live workshops that I run with my colleagues run throughout the year.

You can contact me at:
info@the10-secondsmathsexpert.com to register your interest in attending my workshops or future publications of my book.

Finally, I want to wish you all the success and have fun. I would love to meet you in person one day in the near future.

Start your journey of discovery...............you are amazing..........go after your dreams and goals.

"SUCCESS IS A JOURNEY – NOT A DESTINATION"

All the success and very best wishes

Philip Chan

The 10-Seconds Maths Expert

ABOUT THE AUTHOR

Philip Chan has been a successful teacher for over forty years and he has extensive experience in teaching at all levels of expertise. This ranges from Primary School level through to High Schools, plus College Students, as well as Adult Education. Philip has conducted hundreds of mathematics workshops, working with children and their parents together. Over a number of years using his vast skills mentoring trainee teachers to empower them in the classroom and workshops using his fun and unique techniques, to create excitement and confidence in an instant.

Many of his students have successfully gone on to some of the leading UK universities such as Oxford and Cambridge to gain their PhD and First Class Honours Degrees, as well as a number of his student becoming top leaders and appointed to senior positions in both business and education.

Philip is a qualified Life Coach and NLP Practitioner working with groups and individuals on personal development. Philip is a former Elite Sports Performance Coach and has helped many athletes progress to competing at National, International and Olympic standards.

He has been successfully working and mentoring some of the top Executives from UK Blue Chip companies and helped several Global Billion Dollar companies with their expansion plans and development over a number of years.

For more than forty years, Philip has been involved in fundraising for a number of charities including UNICEF,

Shelter, Oxfam, YMCA and many others by giving informative talks on subjects like: Stress Management; Prevention and Recovery from serious illnesses, such as cancer, without the use of drugs. Other talks include Relaxation Techniques and Memory Training in preparation for academic examinations. All donations go directly to the charities concerned.

Philip is also a motivational speaker and has enriched the lives of countless people in achieving their goals and dreams. Currently he is working with a number of the world's top business coaches from the UK, Canada, USA, South Africa and Australia plus other countries to develop businesses for mutual benefit.

He is a member of the Professional Speakers Academy both as a speaker and a trainer to coach people to stand out and how to deliver their unique message in a way in line with their values to make a difference.

His goal and dream is to inspire the next generations to develop a positive attitude for learning in all subjects and empower them in the joy of learning, discovery and raising their self beliefs for greater achievement.

His goal is to share these techniques with children and empower teachers across the world with

'The 10-Seconds Speed Maths Techniques' not only in mathematics but a number of area in the empowerment of the individual to develop their potentials and he would love you to be a part of this success to SHARE THE KNOWLEDGE.

PERSONAL NOTES (MY KEY LEARNING)

PERSONAL NOTES (MY KEY LEARNING)

PERSONAL NOTES (MY KEY LEARNING)

My Affirmation

" I DO MATHS WITH EASE AND I HAVE FUN"

(Say this three times in the morning and three times before you go to bed.)

I wish you all the success in the world.

Best wishes

Philip Chan
The 10-Seconds Maths Expert

PERSONAL NOTES

PERSONAL NOTES

PERSONAL NOTES

18418734R00064

Printed in Great Britain
by Amazon